NATIONAL WATER RESOURCES CHALLENGES FACING THE U.S. ARMY CORPS OF ENGINEERS

Committee on U.S. Army Corps of Engineers
Water Resources Science, Engineering, and Planning

Water Science and Technology Board

Division on Earth and Life Studies

NATIONAL RESEARCH COUNCIL
OF THE NATIONAL ACADEMIES

THE NATIONAL ACADEMIES PRESS
Washington, D.C.
www.nap.edu

THE NATIONAL ACADEMIES PRESS 500 Fifth Street, N.W. Washington, DC 20001

NOTICE: The project that is the subject of this report was approved by the Governing Board of the National Research Council, whose members are drawn from the councils of the National Academy of Sciences, the National Academy of Engineering, and the Institute of Medicine. The members of the panel responsible for the report were chosen for their special competences and with regard for appropriate balance.

This report is based upon work supported by the U.S. Army Corps of Engineers, Humphreys Engineer Center Support Activity under Contract No. W912HQ-09-C-0041. Any opinions, findings, conclusions, or recommendations expressed in this publication are those of the author(s) and do not necessarily reflect the views of the U.S. Army Corps of Engineers, Humphreys Engineer Center Support Activity.

International Standard Book Number-13: 978-0-309-21832-0
International Standard Book Number-10: 0-309-21832-4

Additional copies of this report are available from the National Academies Press, 500 Fifth Street, N.W., Lockbox 285, Washington, DC 20055; (800) 624-6242 or (202) 334-3313 (in the Washington metropolitan area); Internet, http://www.nap.edu.

Printed in the United States of America

THE NATIONAL ACADEMIES

Advisers to the Nation on Science, Engineering, and Medicine

The **National Academy of Sciences** is a private, nonprofit, self-perpetuating society of distinguished scholars engaged in scientific and engineering research, dedicated to the furtherance of science and technology and to their use for the general welfare. Upon the authority of the charter granted to it by the Congress in 1863, the Academy has a mandate that requires it to advise the federal government on scientific and technical matters. Dr. Ralph J. Cicerone is president of the National Academy of Sciences.

The **National Academy of Engineering** was established in 1964, under the charter of the National Academy of Sciences, as a parallel organization of outstanding engineers. It is autonomous in its administration and in the selection of its members, sharing with the National Academy of Sciences the responsibility for advising the federal government. The National Academy of Engineering also sponsors engineering programs aimed at meeting national needs, encourages education and research, and recognizes the superior achievements of engineers. Dr. Charles M. Vest is president of the National Academy of Engineering.

The **Institute of Medicine** was established in 1970 by the National Academy of Sciences to secure the services of eminent members of appropriate professions in the examination of policy matters pertaining to the health of the public. The Institute acts under the responsibility given to the National Academy of Sciences by its congressional charter to be an adviser to the federal government and, upon its own initiative, to identify issues of medical care, research, and education. Dr. Harvey V. Fineberg is president of the Institute of Medicine.

The **National Research Council** was organized by the National Academy of Sciences in 1916 to associate the broad community of science and technology with the Academy's purposes of furthering knowledge and advising the federal government. Functioning in accordance with general policies determined by the Academy, the Council has become the principal operating agency of both the National Academy of Sciences and the National Academy of Engineering in providing services to the government, the public, and the scientific and engineering communities. The Council is administered jointly by both Academies and the Institute of Medicine. Dr. Ralph J. Cicerone and Dr. Charles M. Vest are chair and vice chair, respectively, of the National Research Council.

www.national-academies.org

COMMITTEE ON U.S. ARMY CORPS OF ENGINEERS WATER RESOURCES, SCIENCE, ENGINEERING, AND PLANNING

DAVID A. DZOMBAK, *Chair*, Carnegie Mellon University, Pittsburgh, Pennsylvania
PATRICK A. ATKINS, ALCOA (ret.), Pittsburgh, Pennsylvania
GREGORY B. BAECHER, University of Maryland, College Park
LINDA K. BLUM, University of Virginia, Charlottesville
ROBERT A. DALRYMPLE, Johns Hopkins University, Baltimore, Maryland
MISGANAW DEMISSIE, Illinois State Water Survey, Champaign
TERRANCE FULP, U.S. Bureau of Reclamation, Boulder City, Nevada
LARRY A. LARSON, Association of State Floodplain Managers, Madison, Wisconsin
DIANE M. MCKNIGHT, University of Colorado, Boulder
J. WALTER MILON, University of Central Florida, Orlando
A. DAN TARLOCK, Chicago-Kent College of Law, Chicago, Illinois
PETER R. WILCOCK, Johns Hopkins University, Baltimore, Maryland

NRC Staff

JEFFREY JACOBS, Study Director
SARAH E. BRENNAN, Senior Program Assistant

Acknowledgment of Reviewers

This report was authored by the National Research Council *Committee on U.S. Army Corps of Engineers Water Resources Science, Engineering, and Planning*. A draft version of the committee's report was reviewed by individuals chosen for their diverse perspectives and technical expertise in accordance with the procedures approved by the NRC's Report Review Committee. The purpose of this independent review is to provide candid and critical comments that will assist the NRC in making its published report as sound as possible, and to ensure that the report meets NRC institutional standards for objectivity, evidence, and responsiveness to the study charge. The review comments and draft manuscript remain confidential to protect the integrity of the deliberative process.

We wish to thank the following individuals for their reviews of this report: Richard A. Conway (retired), Union Carbide Corporation; Gerald E. Galloway, University of Maryland; Peter Gleick, Pacific Institute for Studies in Development, Environment, and Security; George R. Hallberg, The Cadmus Group; Chris T. Hendrickson, Carnegie Mellon University; W. Allen Marr, Geocomp Corporation; Denise J. Reed, University of New Orleans; and Marylynn V. Yates, University of California, Riverside.

Although these reviewers provided constructive comments and suggestions, they were not asked to endorse the report's conclusions and recommendations, nor did they see the final draft of the report before its release. The review of this report was overseen by Leo M. Eisel (retired), Brown and Caldwell, Denver, Colo., who was appointed by the National

Research Council. Dr. Eisel was responsible for ensuring that an independent examination of this report was conducted in accordance with NRC institutional procedures and that all review comments received full consideration. Responsibility for this report's final contents rests entirely with the authoring committee and the NRC.

Contents

Summary

The U.S. Army Corps of Engineers (Corps) is responsible for planning, construction, operations, and maintenance of much of the nation's water resources infrastructure. This infrastructure includes flood control levees, multi-purpose dams, locks, navigation channels, port and harbor facilities, and beach protection infrastructure. The Corps of Engineers also regulates the dredging and filling of wetlands subject to federal jurisdictions. Along with its programs for flood damage reduction and support of commercial navigation, ecosystem restoration was added as a primary Corps mission area in 1996.

The National Research Council (NRC) *Committee on U.S. Army Corps of Engineers on Water Resources Science, Engineering, and Planning* was convened by the NRC at the request of the Corps of Engineers to provide independent advice to the Corps on an array of strategic and planning issues (the committee's full statement of task is presented in Box 1; see Appendix C for a listing of and biographical information on committee members). This activity initially will extend over 5 years, during which the committee will issue a report each year. This is the committee's first report.

This report presents several findings, but no recommendations, to the Corps of Engineers based on initial investigations and discussions with Corps leadership. It is intended to serve as a survey of the key water resources challenges facing the Corps, the limits of what might be expected today from the Corps, and future prospects for the agency. The audience for the report includes not only the Corps of Engineers, but also the U.S. Congress, the administration, Corps project co-sponsors, and the many other entities that are affected by Corps projects and interact with the agency. The report will serve as a foundational document to be referenced in the committee's future reports.

The report's findings are as follows:

- **In an earlier era of national water development, Corps of Engineers civil works projects focused on construction of dams, levees, navigation channels, and other infrastructure. Over time, Congress has greatly broadened the Corps' work program and responsibilities. Future Corps water resources activities will be less dedicated to construction of major new civil works, and more heavily focused on (1) operating, maintaining, rehabilitating, and upgrading existing infrastructure, (2) re-allocating reservoir storage and releases among changing water resources demands and users, and (3) providing some degree of ecosystem restoration and ecological services in heavily altered riparian and aquatic ecosystems.**
- **There has been a declining level of investment in the civil works infrastructure owned and operated by the Corps of Engineers. Deferred costs for maintaining the nation's infrastructure for flood and hurricane protection, and for commercial navigation, are considerable.**
- **Despite decreasing emphasis on new construction, Congress and the nation will continue to rely upon the Corps for emergency response activities and for periodic upgrades to civil works infrastructure.**
- **Despite declining investment levels and numbers of Corps personnel, the nation expects the Corps to provide a number of services, including flood risk management, water-based recreation, commercial navigation, ecosystem restoration, hydropower production, water supply, and coastal and beach protection. This situation leads to expectations that the Corps of Engineers and its civil works construction program cannot meet consistently.**
- **The backlog of authorized federal water resources projects that have not yet received appropriations, or which have begun some level of planning or construction and await additional funds for completion, is considerable. There is also a considerable backlog of existing water project and infrastructure maintenance. The collective backlog of unfinished work leads to projects being delayed, conducted in a stop-start manner, and to overall inefficient project delivery.**
- **The modern context for water resources management involves smaller budgets, cost sharing, an expanded range of objectives, and inclusion of more public and private stakeholders in management decisions. Two important implications of these conditions are (1) given current budget realities, the nation may have to consider more flexible,**

innovative, and lower cost solutions to achieving water-related objectives, and (2) the Corps of Engineers will by necessity work in settings with more collaboration and public and private partnerships than in the past.

- The Corps of Engineers is increasingly challenged to provide a wide variety of water project benefits, some of which often are not consistent and compatible with one another. Some of these challenges relate to inconsistencies in authorizing legislation and related regulations, while some relate to the natural limits of hydrologic and ecologic systems. As a result of these factors, the Corps increasingly finds itself involved in controversies over shared water resources that are beyond the agency's mandates and capacities to fully resolve.
- The Corps of Engineers reflects a national water planning paradox: national water resources demands are increasing and becoming more complex, while at the same time, national investments in water infrastructure exhibit a declining trend. Moreover, in some parts of the nation there are additional water management objectives relevant to Corps project operations, such as water quality goals, in which the agency may be requested to expand its involvement.
- The nation's water planning needs and priorities promise to become even more contentious, complex, and harder to anticipate, in the future. As this report describes, the nature of water planning and the typical water resources project have shifted over the past 50 years. Examples of topics of priority and concern across the United States today are improved flood risk management, efficient water infrastructure planning and investments, navigation infrastructure, water-based recreation, hydropower generation, water supply, ecosystem restoration and endangered species protection, water quality, and understanding and managing water-related risks associated with hydrologic nonstationarity. The Corps of Engineers' authorities, levels of staffing, and resources have gone through changes over the years. At the same time, the Corps of Engineers retains a clear leadership role in many of the nation's major river and aquatic systems, and there will be a continued need for an innovative and responsive Corps of Engineers to lead efforts in addressing national water planning challenges.

National Water Resources Challenges Facing the U.S. Army Corps of Engineers

INTRODUCTION

The U.S. Army Corps of Engineers has a long history of planning and managing important aspects of the nation's water resources, including flood hazards, commercial navigation, coastal protection, and more recently, ecosystem protection and restoration. At the direction of the U.S. Congress, the Corps of Engineers has constructed, and today operates and maintains, multi-purpose dams, navigation channels, and levees in all fifty U.S. states. The Corps also plays key roles in port and harbor maintenance and in dredging and waterways management.

For much of its history, the Corps has enjoyed considerable authority and public confidence. Corps of Engineers dams, locks, navigation channels, levees, and other infrastructure have conferred many benefits to many parties and the nation. The Corps also has been a recognized leader in hydrologic and hydraulic engineering theory and practice, both nationally and around the world.

Over the years, changes in Corps decision-making authority, federal budgetary priorities, and shifts in national water planning goals have affected the agency's water resources management program. Federal environmental legislation, emerging demands for protection and restoration of aquatic ecosystems, declining levels of federal investments in civil works infrastructure, devolving authority, and a need to provide benefits for an expanding range of constituents and sectors have increased the complexity and controversies surrounding the operational decisions, planning studies, and water resources work of the Corps. The expertise and decision making of the Corps has been challenged in many instances and some of the public confidence the agency enjoyed in an earlier era has eroded.

This report is from the National Research Council (NRC) *Committee on U.S. Army Corps of Engineers Water Resources Science, Engineering, and Planning* (see Appendix C for a listing of and biographical information on committee members). The committee was established in late 2009 with sponsorship from the Corps of Engineers. The committee's mandate is to provide strategic advice on emerging water resources issues and challenges (see Box 1 for this committee's full statement of task). This initial 5-year project calls for annual reports. This is the first report in that series.

The Corps of Engineers is an agency within the U.S. Department of Defense and has both civilian and military responsibilities. Under the Corps civil works program and at the direction of Congress, the Corps plans, constructs, operates, and maintains a wide range of water resources projects. The Corps' military program provides engineering, construction, and environmental management services for Department of Defense agencies (Carter and Cody, 2006). This report focuses on the

BOX 1
Statement of Task
Committee on U.S. Army Corps of Engineers Water Resources Science, Engineering, and Planning

This committee will provide advice to the Corps of Engineers on a range of scientific, engineering, and water resources planning issues through periodic reports. This committee's first report will identify emerging national water resources challenges and their implications for Corps of Engineers strategies and programs. The statements of task for subsequent reports will be determined through discussions between the committee and the Corps, and will be subject to approval of the NRC Governing Board's Executive Committee.

Through its reports, the committee will provide advice to the Corps on agency practices that are valid or that should be revised, and help the Corps anticipate and prepare for emerging water resources planning challenges. Meetings between this committee and the Corps will allow for the identification of important and emerging water resources planning and policy issues of high priority to the agency and upon which they are seeking external advice. In addition to speaking with the Corps of Engineers, the committee often will engage invited speakers from other federal agencies, U.S. congressional staff, state governments, the private sector, and relevant stakeholders. The committee also may serve as a forum for occasional workshops on thematic issues, such as flood risk management, sustainable river system planning, hydroecosystem restoration, or implications to water management of climate change and variability.

Corps of Engineers' water resources activities within its civil works mission. Within the civil works program, this report focuses on Corps water resources and management, and Corps infrastructure and activities that affect river and coastal hydrology and hydraulics. It does not investigate or comment upon the Corps' regulatory activities within its civil works program, as those Corps activities are less related to planning and more related to regulating actions and impacts of other entities. The report also does not investigate Corps military engineering responsibilities and activities.

As explained in Box 1, this committee's first report focuses on identifying "emerging national water resources challenges and their implications for Corps of Engineers strategies and programs." This report addresses several overarching water resources planning issues as they relate to the Corps of Engineers. It is anticipated that these subsequent reports will explore specific Corps of Engineers program areas, special initiatives, and reports, and thus will investigate scientific and engineering issues in greater detail. Topics for the committee's subsequent reports are not yet determined. As this project proceeds, future statements of task for this committee will be negotiated based on discussions between the Corps and the NRC regarding timely and appropriate topics. Preliminary discussions regarding the committee's second report point to a prospect of a review of Corps of Engineers flood risk management programs.

This report was prepared on the basis of two open public meetings, which included presentations from several Corps of Engineers staff and other invited speakers, and a third, closed meeting at which the committee discussed its draft report. The committee has reviewed documents provided by the Corps and various stakeholders prior to and during the public meetings, and has reviewed other pertinent information, including a number of previous NRC reports on U.S. water resources issues and the activities of the Corps. Given the long-term nature of this project, as well as the many factors that affect Corps of Engineers policies and actions, this committee's first report focuses on elucidation of the decision-making context in which the Corps operates. It offers several observations and findings about the conditions that affect and constrain the Corps of Engineers, but no recommendations. This report will serve as a foundational document that the committee will refer to and build upon in subsequent reports.

TWENTY-FIRST-CENTURY WATER RESOURCES PLANNING AND MANAGEMENT

The nature of the Corps of Engineers water resources program and its water projects have changed greatly over the past 50 years. The thrust of hydrologic engineering activities across the nation has moved from an earlier era of building civil works infrastructure to a greater emphasis today on infrastructure maintenance and on restoring aquatic ecosystem functions and services in significantly altered hydro-systems. Dams have been constructed for flood control and/or navigation enhancement on most of the major rivers and many of their tributaries, and the flows and physical characteristics of nearly every major U.S. river are heavily controlled. Many of the nation's largest dams and related hydropower, navigation, and water supply systems were constructed in the mid-twentieth century by the Corps of Engineers and the U.S. Bureau of Reclamation. In the 1930s, for example, Hoover Dam (Reclamation) on the Colorado River, Fort Peck Dam (Corps) on the Missouri River in Montana, and many of the navigation locks and dams on the upper Mississippi River (Corps) and the Ohio River (Corps) were constructed. In the 1940s, Grand Coulee Dam (Reclamation) on the Columbia River and the Shasta Dam (Reclamation) on the Sacramento River were completed. The 1950s and 1960s saw the construction of five major mainstem dams across the Missouri River (Corps), along with Glen Canyon Dam (Reclamation) on the Colorado River. The Corps of Engineers has constructed about 800 dams across the nation for a combination of flood control, hydropower production, navigation enhancement, and water supply purposes (see Appendix B). In addition, the Corps operates and maintains 12,000 miles of commercial inland channels, and maintains over 900 coastal, Great Lakes, and inland harbors. The Corps also has constructed or operates nearly 14,000 miles of levees in the federal levee system (USACE, 2011a).

The Corps of Engineers continues to play important roles in flood risk reduction, commercial navigation, and more recently, ecosystem restoration across the nation and in most of the nation's major riverine, lacustrian, and coastal systems. The Corps, for example, is responsible for developing operating plans for the Missouri River mainstem dam and reservoir system, operations on the Upper Mississippi River-Illinois Waterway system, and is a leading partner in restoration activities in Florida's Greater Everglades ecosystem. The thrust of Corps activities in these, and other, hydrologic and aquatic systems has shifted. Today there is less construction of large civil works projects but the operations and maintenance (O&M) functions for all existing infrastructure remains. These O&M functions today include working with stakeholders in setting reservoir release schedules and navigation pool elevations, ecosystem restoration, and endangered species protection.

The Corps also is challenged to operate in a setting of changing hydrologic realities and demands. For example, increasing urban water demands in some regions of the country, such as in the greater Atlanta metropolitan area, are stressing existing water supplies. At an aggregate national level, however, total water withdrawals have leveled off, and total water withdrawals in 2005 were "slightly less than the estimate for 2000, and about 5 percent less than total withdrawals in the peak year of 1980" (Kenny et al., 2009). The Corps also must cope with hydrologic variations and extremes, owing to changes in land use, climate parameters, or some combination. The concept of hydrologic "nonstationarity" (Milly et al., 2008) is prompting reconsideration of some basic tenets of hydrologic analysis and hydraulic engineering, which will affect Corps planning and operations going forward. Potential impacts of sea level rise (Solomon et al., 2007) on Corps infrastructure and coastal ecosystems present important planning and operational issues. Changing water withdrawal patterns and shifting climate and hydrologic regimes are among many national water resources challenges facing the Corps of Engineers and other federal, state, and local entities with water planning and management responsibilities. (Table 1 provides a list of national water resources planning challenges identified by this committee; also see NRC, 2010, for a similar list.)

In response to these many water planning challenges, the Corps of Engineers has been adjusting its priorities and activities to reflect new objectives and priorities for the use of water resource systems. The Corps' traditional responsibilities were in the areas of navigation enhancement and flood risk reduction. Corps of Engineers multi-purpose dams and

TABLE 1 National Water Resources Challenges

- Integrating floodplain management, risk management, public safety, and ecosystem values;
- Aging water control infrastructure and port and inland navigation facilities;
- Urban stormwater management and water supply;
- Watershed restoration;
- Quantifying and leveraging ecosystem services;
- Accommodating diverse stakeholder preferences in operational decisions;
- Integrating social and cultural values into technical aspects of water project decision making;
- Defining tolerable risk for water resource projects;
- Population and economic growth and increasing water demands;
- Increasing demand for water resource projects with diminishing ability to fund those projects;
- Reallocating water resources to new uses, including ecosystem restoration;
- Mitigating nonpoint source discharges to improve water quality;
- Planning for extreme climate events and changing climate conditions (nonstationarity);
- Protection of endangered species while simultaneously meeting demands of traditional water users.

water projects also provided hydropower, water supplies, and recreational benefits. In the mid-1990s, the Congress authorized the Corps to construct projects for ecosystem restoration goals. Over time the demands from the Corps and its water resources projects have broadened, and the Corps often has been expected—and challenged—to provide benefits to both traditional and newer project beneficiaries.

The Corps of Engineers, and the nation, today are constructing fewer civil works projects for water development than during the 1950s and 1960s. The Corps in many instances is working to provide a wider suite of benefits with existing water resources infrastructure. The Corps often is working in heavily altered ecosystems with major impoundments and extensive navigation channels, and many of their projects entail hydrologic and related adjustments (e.g., sediment transport and deposition processes) to these systems and a re-allocation of water-related benefits.

In an earlier era of national water development, Corps of Engineers civil works projects focused on construction of dam, levees, navigation channels, and other infrastructure. Over time, Congress has greatly broadened the Corps' work program and responsibilities. Future Corps water resources activities will be less dedicated to construction of major new civil works, and more heavily focused on (1) operating, maintaining, rehabilitating, and upgrading existing infrastructure, (2) re-allocating reservoir storage and releases among changing water resources demands and users, and (3) providing some degree of ecosystem restoration and ecological services in heavily altered riparian and aquatic ecosystems.

NATIONAL WATER MANAGEMENT INFRASTRUCTURE AND INVESTMENTS

The nation has made considerable investments in water resources infrastructure in the form of dams, locks, levees, navigation channels, shoreline protection, and port and harbor facilities. As the extent of this infrastructure has grown, investment needs and priorities naturally have shifted from an emphasis on constructing new infrastructure, to operating, maintaining, and upgrading existing infrastructure. At the same time, demands for new and replacement infrastructure will continue. The investments that will be required to maintain and upgrade this infrastructure are considerable.

The past three decades have seen a significant decline in the value of the Corps of Engineers capital stock portfolio. An estimate of the value of that portfolio shows a decline from a peak value of $250 billion in 1983, to $165 billion today (2011; see Figure 1).

The trend of steadily declining investments in the nation's water

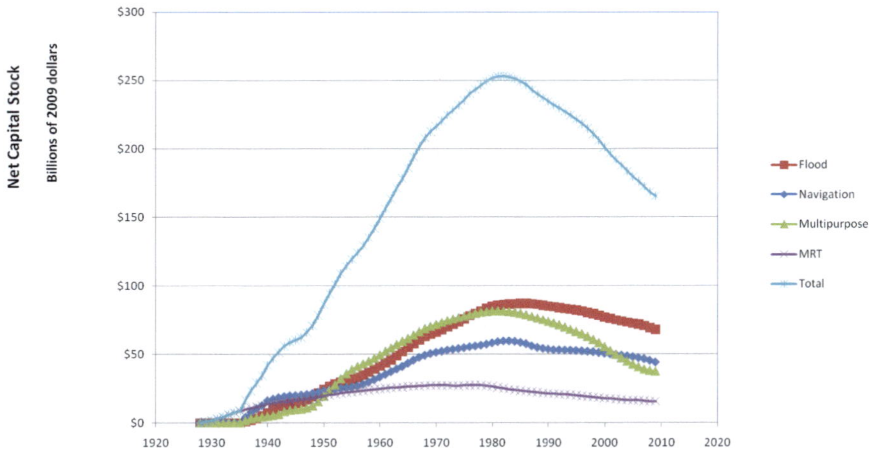

FIGURE 1 Net Capital Stock Estimates of Corps Civil Works Projects, 1928-2009 (2009 dollars). Investments made prior to 1928 are not included in the figure. The MRT acronym refers to "Mississippi River and Tributaries."
SOURCE: Steven Stockton, U.S. Army Corps of Engineers, personal communication, 2010.

resources infrastructure is sobering in many respects. For example, some implications of these declining investments are reflected in the state of the nation's levees. There are approximately 14,000 miles of levees in the federal levee system (USACE, 2011a) that were constructed by and/or are overseen by the Corps (and not counting levees built by local entities such as reclamation and drainage districts), with thousands of flood-prone communities located behind them. Many levees are in a state of deterioration, and many citizens and communities behind these levees have only a limited appreciation of the flood risks to which they are exposed.

There is no national inventory of all these levees, nor is there a systematic program of levee inspection and maintenance. A National Committee on Levee Safety was established in the 2007 Water Resources Development Act to develop a report and strategy for addressing national levee safety challenges. In addition, Congress funded the Corps of Engineers to undertake a national levee inventory, which is now completed, but mainly for levees in the federal levee system (USACE, 2011a). Similarly, many of the nation's locks and dams that support commercial navigation are aging and in need of upgrades and repairs (e.g., see discussion of aging Ohio River locks and dams in Box 2). On the Upper Mississippi River-Illinois Waterway, in Title VIII of the 2007 Water Resources Development Act (WRDA) Congress authorized approximately $2 billion for new locks

BOX 2
Navigation Infrastructure Maintenance and Renewal on the Ohio River

The Ohio River has been changed significantly by an extensive system of 21 lock-dam facilities which allow for pool maintenance and year-round transportation of bulk commercial goods, especially coal. Without these navigation control systems, river transport and riverfront development would be significantly less than they are today. The Ohio River basin region has become dependent on the lock and dam infrastructure and the management services that it makes possible. However, these services are at increasing risk as the infrastructure and equipment that make year-round navigation possible and reliable continue to age without adequate maintenance and renewal.

An example of the infrastructure renewal challenge is provided by the Upper Ohio River navigation system (USACE, 2011c). The Emsworth, Dashields, and Montgomery lock-dam facilities are the first three locks and dams on the Ohio River below Pittsburgh. All three facilities were built prior to 1940 (Emsworth, 1922; Dashields, 1929; Montgomery, 1936), and are the oldest structures with the smallest lock chambers in the Ohio River system. In addition to limitations posed by inadequate lock capacity, all three have serious structural problems due to their age and do not conform to modern design criteria. All three facilities underwent major rehabilitation in the 1980s and early 1990s to extend the useful life of the projects another 25 years, but many known problems were not addressed, the 25-year extension period is now coming to an end, and substantial renewal can no longer be avoided. The Corps of Engineers undertook a long-range planning effort for the Upper Ohio navigation system in 2007 which is scheduled for completion in 2012. If funding for study completion, engineering and design, and construction is seamless and project authorization is timely, then lock replacement and other work at the Emsworth, Dashields, and Montgomery facilities could be potentially completed in 10-15 years. The history of similar projects in the Ohio River basin, however, suggests that financing and construction of new facilities will take much longer.

In the absence of funding for replacement of the infrastructure to return the Ohio River lock and dam system to its original reliability, and to bring the system up to modern design standards and capacity requirements, the three Upper Ohio River locks and dams are kept operating with repairs funded under the annual operating and maintenance budget of the Pittsburgh District of the Corps of Engineers. Funding for improvements to the infrastructure is largely limited to those that are deemed "emergency" in nature, e.g., the replacement of two gates on the Montgomery Dam which were damaged by a runaway barge in 2008. Eight additional gates are in need of replacement at the Montgomery Dam, but funds are not available for the $5 million cost of each gate.

SOURCES: David J. Heidish, U.S. Army Corps of Engineers, Pittsburgh District, personal communication, 2011; USACE, 2011c.

(that would replace existing ones) at five locations on the Mississippi River and two on the Illinois River.

Despite the declining trend in infrastructure investment, the Corps of Engineers can be expected to be directed occasionally to undertake a high-priority civil works project. There is no better example than the Corps' post-Katrina work in strengthening the Greater New Orleans Hurricane and Storm Damage Risk Reduction System. Following Katrina, the U.S. Congress provided $14.5 billion of federal funding for levee and flood-wall repair and strengthening, pump station repairs and upgrades, levee armoring, and other infrastructure improvements across the Greater New Orleans metro area (USACE, 2011b). These construction activities extend across 350 miles of protective structures and include a large surge barrier (Inner Harbor Navigation Canal Surge Barrier). The Corps' post-Katrina efforts illustrate a type of flood risk and emergency response challenge that the Corps will face in the future despite the declining trend in civil works infrastructure investment. These conditions also point to a national need for a Corps of Engineers that is able to respond to not-fully-anticipated water planning or emergency response needs.

There has been a declining level of investment in the civil works infrastructure owned and operated by the Corps of Engineers. Deferred costs of maintenance of the nation's aging flood and hurricane protection, and navigation, infrastructure are considerable. Despite decreasing emphasis on new construction, Congress and the nation will continue to rely upon the Corps for emergency response activities and for periodic upgrades to civil works infrastructure.

CORPS OF ENGINEERS CAPACITIES AND PLANNING

Regardless of the trend in reduced investments in civil works infrastructure, needs and demands continue for operations and maintenance, upgrades in existing infrastructure, and for new water resources projects. At the same time, the Corps of Engineers remains responsible for operations and maintenance of a vast, existing infrastructure that includes approximately 700 dams and nearly 12,000 miles of federal levees (Appendix B). In addition to traditional services of flood control and navigation, the Corps of Engineers is often authorized to provide complementary benefits (e.g., water supply, hydroelectric power production), along with more recently authorized project purposes (e.g., aquatic ecosystem restoration).

In a previous era, the main disciplines represented within Corps of Engineers project planning were engineering, hydrology, hydraulics, and economics. These disciplines remain important in Corps planning and project operations. However, changes in the nature of Corps water projects have entailed changes in the analytical methods and planning approaches needed to address a new planning environment. Examples of scientific and technical fields and applications that are driving the need for "new expertise" include infrastructure sensing; decision support systems; modeling and forecasting of global grain markets; integrated hydrological and ecosystem modeling; conceptual ecological model development for endangered species and other species of interest; waterway traffic modeling; risk analysis and communication; valuation of environmental benefits; and large-scale hydrologic system modeling. Also, as the Corps often plays a prominent role in facilitating multi-objective decisions among competing user groups, concepts and applications such as adaptive management, conflict resolution, and facilitation have become increasingly important.

A challenge for the Corps of Engineers in meeting water planning needs in today's water management era is the need to develop new expertise at a time of a long-term decline in the number of Corps personnel and declining budgets for Corps research. Over the past three decades, the Corps has experienced a declining number of personnel. Employment within its civil works programs has dropped by roughly 25 percent since an employment peak in 1983 (but has experienced a recent uptick; see Figure 2).

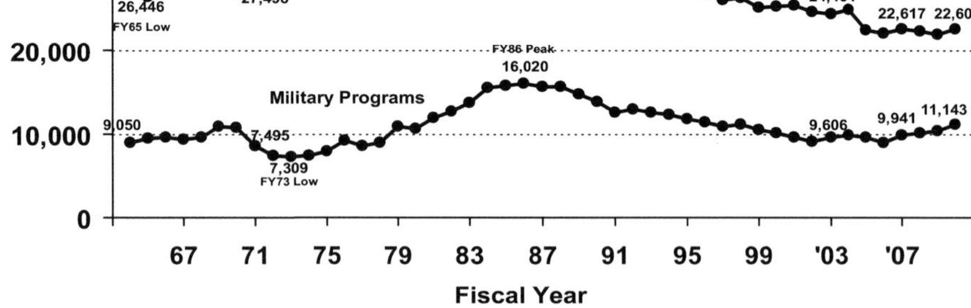

FIGURE 2 Corps of Engineers Staffing Trends, 1964-2010.
SOURCE: Bruce Carlson, U.S. Army Corps of Engineers, personal communication, 2011.

The Corps' ability to adopt and implement more contemporary planning approaches may be stymied further by planning guidance and regulations that have not been updated to adequately reflect modern principles. For instance, the Corps of Engineers is mandated to comply with the federal *Economic and Environmental Principles and Guidelines for Water and Related Land Resources Implementation Studies* (U.S. Water Resources Council, 1983) to guide its water planning studies. Although of great utility in an earlier era, many groups have found the "Principles and Guidelines" document in need of modernization to conform better to today's water planning context and needs (e.g., NRC, 2004). The ongoing effort by the White House Council on Environmental Quality (CEQ) to update the Principles and Guidelines document offers an opportunity to promote more efficient water resources planning by the Corps of Engineers and other federal agencies (see NRC, 2010, for a review of CEQ's initial proposed changes to this document).

Despite declining investment levels and numbers of Corps personnel, the nation expects the Corps to provide a number of services, including flood risk management, water-based recreation, commercial navigation, ecosystem restoration, hydropower production, water supply, and coastal and beach protection. This situation leads to expectations that the Corps of Engineers and its civil works construction program cannot meet consistently.

AUTHORIZATION AND APPROPRIATIONS PROCESSES

Traditional Corps of Engineers water resources projects receive congressional authorization in a federal Water Resources Development Act, and receive project funding through a separate appropriations process. There is a growing number of federal water resources projects that have received congressional authorization, but are awaiting appropriations. This applies both to projects that have not yet started, and to projects that have begun but are unfinished. There also is a considerable backlog of maintenance.

One implication of this "backlog" of authorized projects that are not fully (or at all) funded is that newly authorized projects often must wait for appropriations until projects in the backlog first receive funding. The monetary value of the current backlog is estimated to be nearly $60 billion dollars (see Box 3 for more discussion of this backlog).

The size of this backlog implies that communities or other project beneficiaries often will wait years, if not a decade or longer, to receive federal water project funding. Further, the backlog exists in a setting in which a Corps of Engineers planning study typically requires several years (NRC, 1999). These realities can be frustrating for communities or other project

BOX 3
Corps of Engineers Civil Works Construction Backlog

The term "backlog" is used to describe generally Corps of Engineers water resources projects that have received congressional authorization, and await appropriations. Usually these projects have either been funded for construction, authorized by Congress, or have been identified in a feasibility report and continued into pre-construction engineering and design. The backlog represents the balance to complete construction for these specifically defined projects of known scope that local interests expect the Corps to build.

Projects in the backlog are normally divided into three categories: active, deferred, and inactive projects. Active projects are usually funded and supported by the non-federal sponsor and/or have been authorized. All these projects are being actively pursued.

The backlog for the Active projects is currently $59.6 billion.

Deferred projects have doubtful economic justification and need further study in order to determine their economic feasibility. Generally, they are not opposed by the nonfederal sponsor, but the nonfederal sponsor is unable to provide required cooperation. Inactive projects are not economically justified and it is anticipated that a restudy would not develop a justified plan.

The inactive and deferred projects backlog is about $2 billion. Since it is unlikely that these projects will ever be built, that amount is of little consequence to the overall backlog.

SOURCE: Bruce Carlson, U.S. Army Corps of Engineers, personal communication, 2011.

co-sponsors that perceive a need for federal support in new water project construction, or for upgrades or maintenance of existing infrastructure.

Another implication of this large and increasing backlog is that the demands for federally funded water resources projects are affecting the ability of the federal government, and agencies like the Corps of Engineers, to construct such projects reliably and efficiently (NAPA, 2007). Authorization of a federal water project through the WRDA process does not include a plan or timeline for project appropriations. Even if an authorized water project eventually receives federal appropriations, that funding often is delivered in incremental amounts. This process of partial project funding through the annual appropriations process results in many projects moving forward in a piecemeal, stop-start manner. This state of affairs can result in inefficient project delivery and higher overall costs, and may be damaging the Corps of Engineers' credibility.

Given this backlog and the realities of modern budget constraints, the Congress and the nation may need to consider more efficient, creative, and flexible approaches to managing flood risks, waterway congestion, endangered species and ecosystem protection, and other water-related challenges. The Corps and the nation may need to place a stronger emphasis on approaches and projects with reduced operation and maintenance needs for ensuring their long-term sustainability.

The backlog of authorized federal water resources projects that have not yet received appropriations, or which have begun some level of planning or construction and await additional funds for completion, is considerable. There is also a considerable backlog of existing water project and infrastructure maintenance. The collective backlog of unfinished work leads to projects being delayed, conducted in a stop-start manner, and to overall inefficient project delivery.

DECENTRALIZATION OF PLANNING AND OPERATIONAL DECISIONS

Another trend in U.S. water policy has been some decentralization of decision making, with more shared responsibilities among the Corps, and with project co-sponsors and other stakeholder groups. Although demands for federal involvement in and funding of water projects remain strong, several factors have relegated more decision-making authority to local entities and stakeholders.

Cost sharing. WRDA 1986 increased the financial responsibilities of local project co-sponsors. With increasing fiscal responsibility, local sponsors understandably have demanded a stronger voice in project planning and implementation. Although the Corps of Engineers has sought to become more responsive to local needs, cost-sharing requirements may be driving the Corps away from systems-based, watershed planning (NRC, 1999).

Stakeholder collaboration. Cost sharing and other requirements have resulted in more direct input from local stakeholder groups in water policy planning and decisions (NRC, 1999). The Corps has recognized this and is actively participating in multi-stakeholder forums in many areas of the country, such as the Missouri River Recovery Implementation Committee (MRRIC). The Corps thus is more frequently collaborating with local sponsors and other parties affected by federal water projects in identifying water resources priorities, resource limits, and trade-off decisions (see USACE and Illinois Department of Natural Resources, 2007, for an example of Corps-State of Illinois DNR cooperation). The Corps also is striving to strengthen relations and collaboration with nongovernmental organizations (see The Nature Conservancy, 2011) with interests and programs in water quality and ecosystem restoration. These conditions and trends present opportunities for the Corps to help multiple parties better understand hydrologic and aquatic systems, and work toward compromise and consensus agreements which can be updated as environmental factors and economic conditions and preferences change over time.

Recognition of the limits of "top down" water management. Federal leadership remains important for numerous water programs and functions. At the same time, in fields such as flood risk management and mitigation, the importance of local and state government input and in sharing risks, and factors such as local land use zoning and regulations, have become increasingly clear and recognized. For example, a recent NRC report that advised the Corps of Engineers on hurricane protection and ecosystem restoration in southern Louisiana encouraged the Corps to work more closely with state and local entities. The report noted that, for example, "It is necessary to clearly delineate the roles of federal and state governments in the collaborative design and development of a comprehensive system that includes coastal, structural, and non-structural protections" (NRC, 2009b, p. 38). In recognition of these conditions and trends, the Corps of Engineers has established its Silver Jackets program. The program aims to provide Corps of Engineers technical assistance in flooding and other natural disasters within collaborative partnerships with state and local entities (see USACE, 2011d, for more information on the program).

Budget realities and constraints. Figure 1 reflects declining appropriations for federal water resources projects. The Corps recognizes this trend and acknowledges that its future roles will include more collaboration and partnerships, noting that they will "work collaboratively with a broad range of stakeholders to help solve water resources problems in an integrated and sustainable manner" (USACE, 2010). Moreover, these budget realities may herald a future with a stronger emphasis on projects that employ creative policy and operational strategies, and that entail reduced maintenance requirements and promote long-term sustainability. Examples of water planning and management approaches that may become more prominent and necessary in the future are floodplain management activities such as zoning, flood risk communication, and evacuation planning (see ASFPM, 2007; NRC, 2009a), and greater reliance on ecosystem services provided by restoration of wetlands and historic streambeds.

These changes pose challenges to a traditional, construction-oriented agency like the Corps of Engineers, and they portend a need for the Corps to make some shifts in overall program emphasis and staffing skills and disciplines. At the same time, they present opportunities for the Corps to provide needed services in nontraditional areas such as facilitation, adaptive management, and inter-agency cooperation. A good example of a proactive Corps of Engineers initiative in these areas is the development of a "Shared Vision Planning" approach at the Corps' Institute for Water Resources (USACE, 2009). The computer-based, collaborative planning approach supported by Shared Vision Planning holds promise in supporting the Corps as a facilitator and partner in multiple-stakeholder dialogue and in adaptive management. These areas and water planning approaches will be increasingly important for the Corps and essential to more efficient national water management.

The modern context for water resources management involves smaller budgets, cost sharing, an expanded range of objectives, and inclusion of more public and private stakeholders in management decisions. Two important implications of these conditions are (1) given current budget realities, the nation may have to consider more flexible, innovative, and lower cost solutions to achieving water-related objectives, and (2) the Corps of Engineers will by necessity work in settings with more collaboration and public and private partnerships than in the past.

Decentralization of some planning and operational decisions presents challenges for the Corps, but it also presents opportunities to participate in and facilitate decision making with its partners to leverage resources external to the Corps in new ways.

TRADITIONAL RESPONSIBILITIES AND EXPANDING MISSIONS

The past 50 years have seen the passage of landmark environmental and other legislation which has had important implications for the water resources work of the Corps. Key examples of this legislation are the National Environmental Policy Act of 1969, the Clean Water Act in 1972, and the Endangered Species Act in 1973. This legislation created numerous new responsibilities and requirements for the Corps of Engineers. More recent environmental legislation, such as proposed numeric nutrient criteria for lakes and flowing waters in Florida (USEPA, 2010), is creating additional planning and operational challenges for the Corps. The Corps of Engineers also is constrained and driven by a large body of other laws and authorizations. Like other federal agencies with multiple missions governed by multiple laws and associated regulations, it is a challenge for the Corps to reconcile inconsistencies within this body of laws and regulations which in aggregate comprise national water policy (NRC, 2004).

The new legislation and regulations have also charged the Corps of Engineers to undertake restoration of aquatic and wetland ecosystems in addition to the traditional water resources missions of flood control and navigation infrastructure. The Corps adopted ecosystem restoration as a mission and explicit project purpose in 1996, and restoration projects today comprise some of the agency's largest activities—especially the Florida Everglades and coastal Louisiana restoration projects.

One result of these changes is that the Corps often is in a position of trying to accommodate a range of very different constituencies, and trying to achieve a multitude of objectives that are not consistent and compatible. An example would be a conflict between meeting hydropower production goals, and ecosystem restoration to support endangered species, as each of these missions may require different, perhaps incompatible, reservoir release schedules. The Corps is often in the position of having to make a decision between operating a project for its legally mandated purpose, or complying with environmental protection mandates and ecosystem restoration recommendations from other agencies, the states, or the public. (One example of where the Corps must reconcile potentially conflicting legislation is in the Missouri River mainstem dam and reservoir system; see Box 4.)

The modern context of numerous objectives, constituencies, and governing legislation presents a complex management environment for the Corps. Like other federal agencies with multiple missions governed by a wide variety of often unintegrated and inconsistent laws and regulations, the agency must reconcile its inconsistencies in the absence of a formal national water policy or any body charged with developing and implementing one (Craig, 2008). Furthermore, because of natural limits of water resources systems, many Corps of Engineers' actions and policies today necessarily entail "zero sum"-type trade-off decisions—or decisions which do not create new benefits, but rather re-distribute existing benefits among competing users (also see NRC, 2011).

This context of water policy decisions obviously can be contentious and it poses considerable challenges to the Corps. Reconciling these diverse and often conflicting obligations is at the core of "integrated water management," which can be challenging in practice because of the fragmented, multi-stakeholder, multi-objective, and litigious water resources planning environment. Furthermore, these challenges may be magnified by factors such as scientific uncertainties, inter-agency or interstate disagreements, or unrealistic stakeholder expectations.

The Corps of Engineers is increasingly challenged to provide a wide variety of water project benefits, some of which often are not consistent and compatible with one another. Some of these challenges relate to inconsistencies in authorizing legislation and related regulations, while some relate to the natural limits of hydrologic and ecologic systems. As a result of these factors, the Corps increasingly finds itself involved in controversies over shared water resources that are beyond the agency's mandates and resources to fully resolve.

BOX 4
Reconciling Multiple Authorities Along the Missouri River

The Corps of Engineers is authorized to operate and maintain the six major mainstem dams along the Missouri River. These dams along the Missouri impound the largest reservoir system in North America. Recently, there has been some devolution of decision-making authority for the Missouri River dam and reservoir system, but the Corps of Engineers clearly has retained its role as the river system's "water master."

The Corps faces innumerable challenges in trying to fulfill the authorized purposes of the dam and reservoir system, which include flood control; navigation; water supply; hydroelectric power generation; recreation; and fish and wildlife. In trying to balance these multiple uses of the river system, the Corps must consider various legislative acts and directives, such as the National Environmental Policy Act, the Clean Water Act, and the Endangered Species Act. In doing so, the Corps works closely with other federal agencies (especially the U.S. Fish and Wildlife Service) in making operational decisions for the river and the reservoir system. The Corps also must work closely with the Missouri River states, tribes, and many other stakeholder interests. The Missouri River Recovery Implementation Committee (MRRIC) was established in 2007 as a multiple stakeholder group to facilitate discussion and negotiations among the Corps and these diverse interests along the river.

The challenges of working under multiple authorities, and the inconsistencies that often arise among them, was clearly stated in 2002 by the then-Commanding General of the Corps Northwestern Division, General David Fastabend:

> *The challenge is that the people of the United States have—over time—told us to do many, many things. In the 1930s and 1940s the American people told us to build, operate and maintain the Missouri River mainstem system for multiple project purposes. Since that original mission, the American people have given us additional instructions. In the 1970s they gave us the Endangered Species Act and the National Environmental Policy Act...As you can well imagine, no one was able to "deconflict" the multiple instructions given to the U.S. Army Corps of Engineers. Our guidance is sometimes contradictory and the resolution of those contradictions is extremely problematic* (Fastabend, 2002).

Two important initiatives taken since the early 2000s by the Corps to help reconcile some of these issues are the updating of its Master Manual for the Missouri River Mainstem Reservoir System (2004) and the establishment of the Missouri River Recovery Implementation Committee (2007).

FUTURE PROSPECTS FOR THE CORPS IN NATIONAL WATER RESOURCES MANAGEMENT AND LEADERSHIP

This report depicts a modern water project planning and policy context that has changed markedly from an earlier era of larger budgets, clearer priorities, and greater deference to Corps of Engineers plans and actions. Demands for goods and services provided by Corps projects, and the nation's riverine, wetland, and other aquatic systems, continue to expand and broaden; at the same time, federal budgets for water resources projects are declining. Furthermore, the Corps of Engineers has experienced a marked reduction in its civilian work force from an employment peak in the early 1980s.

The trends and conditions identified in this report present a water resources planning paradox for the Corps of Engineers and the nation. On the one hand, population and economic growth and increasing urban water demands in some areas of the country, occasional severe drought conditions, and new laws and authorities exert demands on the nation's existing water resources and for infrastructure improvements and upgrades. The products and services provided by an entity like the Corps are as important as ever. At the same time, the Corps of Engineers, with its mandate for national water resources planning, is affected by declining levels of investment, decreases in personnel, changes and shifts in the nature of water resources projects, and staffing needs for the skills and capacities required to most effectively address emerging water challenges and related needs.

The Corps of Engineers clearly has entered an era of resource constraints, along with changing demands from the nation's water systems and for Corps planning expertise. At the same time, the agency largely retains a structure with division and district offices across the nation that existed decades ago when the Corps was constructing far more civil works projects. Given the new realities described in this report, the U.S. Congress and the Corps of Engineers may wish to consider if there is a need to re-align or streamline the agency's organizational structure to better fit modern conditions.

The Corps also may wish to carefully consider the future roles of its centers of expertise. There is likely to be a need for greater efficiencies in future Corps planning efforts. For example, a previous National Research Council committee concluded that, "Creating a means for drawing from Corps personnel across district lines and allowing Corps staff from its centers of expertise, such as the Corps Institute for Water Resources . . . and its Waterways Experiment Station . . . would allow the Corps to bring its best minds to bear upon its more complex planning studies" (NRC, 2004).

Another related organizational issue the Corps may wish to consider is the prospect for strengthening its relationships with the private sector through "public-private partnerships." Collaboration with the private sector promises to become increasingly important for the Corps, and the agency may wish to augment those relationships with additional contracts with the private sector to supplement internal skills, explore project financing alternatives, improve communications within project operations, or other opportunities.

It will be crucial that the Corps, the U.S. Congress, and the numerous partners and constituencies of the Corps acknowledge these new realities. All parties will have to consider the key priorities and responsibilities for the future Corps of Engineers. In order to reach more harmonious and more flexible hydro-system operational decisions and regimes, it will be important for the Corps and water user groups to acknowledge the limits of water system benefits (e.g., water supplies, ecosystem goods and services), and the need to distribute limited resources among many, often competing, users.

Although the changing societal and budgetary context for water resources management clearly presents challenges for the Corps, this also offers opportunities to develop and implement new, creative approaches to water management in which more responsibilities are shared with states and local governments, and other federal agencies. Examples of emerging areas and opportunities, and inter-agency collaboration, for the Corps include helping better integrate water quality regulation and management with water resources planning, groundwater storage and retrieval, and urban stormwater management. It also will be important for the Corps to continue to explore and develop different means and techniques for evaluating and explaining multi-objective, trade-off decisions. Examples of such techniques include negotiation and conflict resolution skills, multi-criteria decision analysis, and resource valuation and monetization.

The Corps of Engineers reflects a national water planning paradox: national water resources demands are increasing and becoming more complex, while at the same time, national investments in water infrastructure exhibit a declining trend. Moreover, in some parts of the nation there are additional water management objectives relevant to Corps project operations, such as water quality goals, in which the agency may be requested to expand its involvement.

The nation's water planning needs and priorities promise to become even more contentious, complex, and harder to anticipate, in the future. As this report describes, the nature of water planning and the typical water resources project have shifted over the past 50 years. Examples of topics of priority and concern across the United States today are improved flood risk management, efficient water infrastructure planning and investments, navigation infrastructure, water-based recreation, hydropower generation, water supply, ecosystem restoration and endangered species protection, water quality, and understanding and managing water-related risks associated with hydrologic nonstationarity. The Corps of Engineers' authorities, levels of staffing, and resources have gone through changes over the years. At the same time, the Corps of Engineers retains a clear leadership role in many of the nation's major river and aquatic systems, and there will be a continued need for an innovative and responsive Corps of Engineers to lead efforts in addressing national water planning challenges.

References

ASFPM (Association of State Floodplain Managers). 2007. National Flood Policy Challenges. Levees: The Double-Edged Sword. Madison, WI: Association of State Floodplain Managers.

Carter, N., and B. Cody. 2006. The Civil Works Program of the Army Corps of Engineers: A Primer. Congressional Research Service Report for Congress. Available online at: http://www.nationalaglawcenter.org/assets/crs/RS20866.pdf.

Craig, R.K. 2008. Climate Change, Regulatory Fragmentation, and Water Triage. 79 University of Colorado Law Review 825, 901.

Fastabend, D.A. 2002. The Army Corps' Missouri River Mission. Water Current Newsletter December 34 (6): 4-8.

Kenny, J.F., N.L. Barber, S.S. Hutson, K.S. Linsey, J.K. Lovelace, and M.A. Maupin. 2009. Estimated use of water in the United States in 2005: U.S. Geological Survey Circular 1344, 52 pp.

Milly, P.C.D., J. Betancourt, M. Falkinmark, R.M. Hirsch, Z.W. Kungzewicz, D.P. Lettenmaier, and R.J. Stouffer. 2008. Stationarity is Dead: Whither Water Management? Science 319: 573-574.

NAPA (National Academy of Public Administration). 2007. Prioritizing America's Water Resources Investments: Budget Reform for Civil Works Construction Projects at the U.S. Army Corps of Engineers. Washington, DC: National Academy of Public Administration.

NRC (National Research Council). 1999. Future Directions in Water Resources Planning for the U.S. Army Corps of Engineers. Washington, DC: National Academy Press.

NRC. 2004. U.S. Army Corps of Engineers Water Resources Planning: A New Opportunity for Service. Washington, DC: The National Academies Press.

NRC. 2009a. The New Orleans Hurricane Protection System: Assessing pre-Katrina Vulnerability and Improving Mitigation and Preparedness. Washington, DC: The National Academies Press.

NRC. 2009b. Final Report from the NRC Committee on the Review of the Louisiana Coastal Protection and Restoration (LACPR) Program. Washington, DC: The National Academies Press.

NRC. 2010. A Review of the Proposed Revisions to the Federal Principles and Guidelines Water Resources Planning Document. Washington, DC: The National Academies Press.

NRC. 2011. Missouri River Planning: Recognizing and Incorporating Sediment Management. Washington, DC: The National Academies Press.

Solomon, S.D., D. Qin, M. Manning, Z. Chen, M. Marquis, K.B. Averyt, M. Tignor, and H.L. Miller (eds.). 2007. Contribution of Working Group I to the Fourth Assessment Report of the Intergovernmental Panel on Climate Change. Cambridge University Press: Cambridge, United Kingdom and New York, NY, USA.

The Nature Conservancy. 2011. Sustainable Rivers Project. Available online at: http://www.nature.org/initiatives/freshwater/partnership/.

USACE (U.S. Army Corps of Engineers). 2009. Shared Vision Planning. Available online at: http://sharedvisionplanning.us/.

USACE. 2010. Watershed Plans. Corps of Engineers EC 1105-2-411. Washington, DC: U.S. Army Corps of Engineers.

USACE. 2011a. The Corps of Engineers Levee Safety Program. Webinar 2: Routine Activities. February 2-3, 2011. U.S. Army Corps of Engineers powerpoint presentation.

USACE. 2011b. Greater New Orleans Hurricane and Storm Damage Risk Reduction System Facts and Figures. Available online at: http://www.mvn.usace.army.mil/hps2/pdf/Jan_Facts_web_2011.pdf.

USACE. 2011c. Upper Ohio Navigation Study. Available online at: http://www.lrp.usace.army.mil/pm/upper_ohio.htm.

USACE. 2011d. Silver Jackets: Many Agencies One Solution. Available online at: http://www.nfrmp.us/state/index.cfm

USACE and Illinois Department of Natural Resources. 2007. Illinois River Basin Restoration Comprehensive Plan with Integrated Environmental Assessment. Available online at: http:www.mrv.usace.army.mil/ILRiverEco/default.htm.

USEPA (U.S. Environmental Protection Agency). 2010. Water Quality Standards for the State of Florida's Lakes and Flowing Waters; Final Rule. Federal Register: 40 C.F.R. Part 131: 75762-75807.

U.S. Water Resources Council. 1983. Economic and Environmental Principles and Guidelines for Water and Related Land Resources Implementation Studies. Washington, DC: U.S. Government Printing Office.

Appendix A

Guest Speakers at Committee Meetings

Washington, D.C. – February 2010

U.S. Army Corps of Engineers
Maj. General William T. Grisoli
Deputy Commanding General for Civil and Emergency Operations

Steven L. Stockton
Director, Civil Works

Sacramento – July 2010

Federal Agencies
U.S. Army Corps of Engineers
Paige Caldwell
Miki Fujitsobo
Alicia Kirchner
Brooke Schlenker
Sara Schultz
Dan Tibbits

U.S. Bureau of Reclamation
Ron Milligan

U.S. Fish and Wildlife Service
Michael Hoover

State Agencies
State of California Department of Water Resources
Rod Mayer
Mike Mierzwa

Levee and Flood Control Associations
Lewis Bair, Reclamation District 108, Grimes, CA
Mike Hardesty, Central Valley Flood Control Association, Dixon, CA

Other Invited Experts
David Ford, David Ford Consulting Engineers, Inc., Sacramento
Ronald Stork, Friends of the River, Sacramento

Appendix B

Corps of Engineers Civil Works Program Statistics[1]

FACTS:

1. PEOPLE

— Civilian employee work years (FTE's), FY 2011: **23,232**
— Military personnel authorized: **294**

2. DIVISIONS & DISTRICTS:

— Number of division offices with Civil Works mission: **8**
— Number of district offices: **38**

3. FUNDING:

Fiscal Year 2010 appropriations: **$5.657 billion**
— Regular Appropriation: **$5.440 billion**
— Supplemental Appropriations: **$217 million**
- — Construction: **$2.031 billion**
- — Operation and Maintenance: **$2.573 billion**
- — Mississippi River and Tributaries: **$359 million**
- — Investigations (e.g. new project studies): **$165 million**
- — Regulatory Program: **$190 million**
- — Formerly Used Sites Remedial Action Pgm. (FUSRAP radiological environmental cleanup): **$134 million**
- — Flood Control and Coastal Emergencies: **$20 million**
- — Expenses and Other: **$185 million**

[1] Statistics are for September 30, 2010, unless otherwise specified

— Other Revenue (estimated)
  — Non-federal (cost-sharing—estimated): **$893 million**
  — Coastal Wetlands Restoration Trust: **$85 million**
  — Permanent appropriation: **$17 million**
— Total program: **$6.652 billion**

4. FUNDING BY BUSINESS LINE, FY 2010 appropriations:
  — Navigation: **$1.822 billion (32.1%)**
  — Flood Risk Management: **$1.942 billion (34.2%)**
  — Environmental (Including FUSRAP& Infrastructure): **$984 million (17.4%)**
  — Regulatory Programs: **$190 million (3.3%)**
  — Hydropower: **$211 million (3.7%)**
  — Recreation: $**284 million (5%)**
  — Emergency Management: **$34 million (0.6%)**
  — Water Supply: **$5 million (0.1%)**
  — Executive Direction & Other: **$185 million (3.3%)**

5. APPROPRIATIONS FOR CIVIL WORKS, PAST 50 YEARS (FY 1961-2010): **$176,370,623,000**
— Adjusted for inflation to Sep 2010: **$358,473,303,000**

6. PROJECTS UNDER CONSTRUCTION, FY 10: **1,167**
— Specifically authorized by Congress: **434**
  — Flood Risk Management: **191**
  — Hydropower: **5**
  — Navigation: **147**
  — Environmental (Including FUSRAP): **39**
  — Environmental Infrastructure: **52**
— "Continuing Authorities" Projects: **733** (Nine authorities, including environmental)

7. FUNDS OBLIGATED, FY 2010 (Current program and prior year funding carryover): **$11,584,480,700**

8. CONTRACTS LET, FY09: **$9.07 billion**
  — To Small Businesses**: $3.87 billion (42.7%)**
  — Small Disadvantaged Firms: **$973 million (10.7%)**

9. DAMS owned/operated by Corps (all purposes) **692**
  — Dams built by Corps but operated by others: **103**
  — Tallest dam: **Dworshak Dam, North Fork Clearwater River, ID, 717 ft.**

10. REAL ESTATE
— USACE owns **136,000** land tracts, totaling more than **7.6 million acres (~11,875 square miles)**
— USACE manages another **4.1 million acres (~6,400 square miles)**
— Total lake surface area at full pool: **26.25 million acres (41,015 square miles–area slightly larger than Kentucky)**
— Largest lake: **Lake Oahe, ND & SD, 587.5 square miles**

11. NAVIGATION
— States served by Corps ports & waterways: **41 (including all States east of Mississippi River)**
— Commercial inland channels operated/ maintained: **12,000 miles**
— Percentage of U.S. domestic freight carried by water (by ton-miles, 2007, excluding air & pipeline): **16%**
— Navigation lock chambers: **238** at **192** sites
  — Locks chambers in operation over 50 years old: **138;** Average age of locks: **58 years**
  — Combined lift of all Corps locks: **6,498 ft.**
  — Highest: **John Day Lock, Columbia R., OR, 110 ft.**
  — Most cargo moved: **Ohio River Lock #52, 80 million tons** (2009)
— Coastal, Great Lakes and inland harbors maintained by Corps: **926**
  — Harbors handling over 250,000 tons of cargo: **183 (111** coastal, **46** Great Lakes, **26** inland) (2009)
  — Port handling most cargo: **South Louisiana, 212.6 million tons** (2009)
  — Value of foreign commerce handled at ports (2009): **$1.156 trillion**
— Tonnage handled by U.S. ports and waterways (2009): **2,211 million tons**
  — Inbound foreign: **858.9 million tons**, Outbound foreign: 494.8 **million tons**, Domestic: **857.1 million tons**
  — Major commodities: **Crude oil, 515.3 million tons; petroleum products, 501.1 million tons; coal & coke, 290.9 million tons; food & farm products, 279 million tons**

12. DREDGING
— Material dredged (construction and maintenance, 2009): **263.6 million cubic yards—enough to fill a football field to a depth of 12 miles**
— Cost: **$1,344.1 million.** Average cost per cubic yard: **$5.10**
— Percentage of material dredged by private firms: **82.2%**

— Companies dredging for Corps: **51** (**33** small businesses) submitted **363** bids for **183** contracts (**87** of which went to small & emerging businesses)
— Percentage of dredging funds going to contractors: **89.2%**
— Corps-owned dredges: **11** (**4** hopper, **7** other)

13. FLOOD RISK MANAGEMENT
— Dams managed by Corps: **692,** at **557** dam projects
— Federal levees built or controlled by Corps: ~**11,750 miles**
— Damages prevented by Corps projects, 2009: **$29.5 billion**
— Average annual damages prevented by Corps projects (2000-2009): **$22.3 billion**
— Damages prevented per $1 invested (adjusted for inflation), 1928-2009: **$7.17**

14. ENVIRONMENTAL PROTECTION/RESTORATION
— Largest projects ($20M+ in FY10):
South Florida Ecosystem Restoration
Columbia River Fish & Wildlife Mitigation,
Missouri River Fish & Wildlife Mitigation
Louisiana Coastal Area Ecosystem Restoration (investigation)
Upper Mississippi River Restoration

15. REGULATORY PROGRAM
— Final Actions, FY10**: 68,800**
— Standard Permits and Letters of Permission: **3,700**
— Activities covered by Regional General Permits: **13,470**
— Covered by Programmatic General Permits: **6,900**
— Covered by Nationwide Permits: **31,900**
— Permits Denied: **275**
— Permits Modified: **3,100**
— Applications Withdrawn: **10,200**
— "No Permit Required" Determinations: **9,810**
— Percent of minor permits completed within 60 days: **92%**
— Jurisdictional Determinations: **63,100**
— Number of approved mitigation banks: **665**
— Compliance visits done on **17%** of mitigation sites and **34%** of mitigation banks or In Lieu Fee sites

16. HYDROPOWER
— Number of projects in operation: **75**, with **350** generating units
— Installed generating capacity: **23,764** megawatts
— Largest USACE power plants:

  - — Capacity—**2,484 megawatts, John Day Dam, Columbia River, OR;**
  - — Most units: **27, Chief Joseph Dam, Columbia River, WA**
  - — Largest generating unit: **220 megawatts, Dworshak Dam, North Fork Clearwater River, ID**

— Annual power generation: **68** billion kilowatt-hours
— Annual gross revenue generated: **approx. $4 billion**
— Repayment to U.S. Treasury from power sales (estimate): **$800 million**
— Value of Hydropower Assets (2007): **approx. $20 billion**
— Rank among U.S. hydropower producers: **#1**
— USACE owns & operates **24%** of U.S. hydropower capacity, or **3%** of total U.S. electric capacity
— FERC licensed non-federal power plants at Corps facilities (not counted above): **90**, with **2,300** megawatts capacity

17. RECREATION
— Rank among Federal providers of Outdoor Recreation: **#1**
— Visits per year: **370 million**
— **10%** of U.S. population visits a Corps project at least once each year
— Number of sites: **4,254** at **422** Corps projects (mostly lakes)
  - — **more than 90% of the lakes are near metropolitan areas (within 50 miles of a MSA)**

— Land & water used for recreation: **12 million acres**
  - — USACE hosts **20%** of visits to Federal recreation areas on **2%** of Federal lands

— Miles of shoreline: **54,879**
— Number of campsites: **92,674**
— Miles of trails: **6,864**
— Number of boat launch ramps: **3,603**
— Share of all U.S. freshwater lake fishing: **33%**
  - — **20,000** fishing tournaments a year

— Spent by visitors at Corps projects: **$18 billion**
  - — Jobs (full or part time) supported by visitation: **350,000**

— Concessionaires on Corps projects: **500**, with gross fixed assets of **$1 billion**
— Volunteers at Corps projects: **54,917**; Hours worked: **1.4 million,** Value of their labor: **$28.3 million**

18. WATER SUPPLY
— Total capacity of Corps lakes: **329.2 million acre-feet**
— Total authorized municipal & industrial water supply storage: **9.76 million acre-feet**

— Total investment associated with municipal & industrial water supply storage: **$1.5 billion**
— Projects with authorized municipal & industrial water supply storage: **136, in 25 States plus Puerto Rico**
— Projects with authorized irrigation storage: **48**

19. EMERGENCY OPERATIONS
— Major disasters responded to in 2010: **20**
    — Largest events: **Hurricane Earl; Flooding in Nashville, Midwest, Arizona and California; Deepwater Horizon Oil Spill; Haiti Earthquake.**
— Corps members deployed to emergency operations: **855**

20. SUPPORT TO OTHER (NON-DEFENSE) GOVERNMENT AGENCIES:
— Number of Federal agencies supported: **70+**
— Expenditures for FY10: **$2 billion**
— Biggest Customers:
    — Dept. of State, $ 630 million
    — Dept. of Veterans Affairs, $348.7 million
    — Environmental Protection Agency, $308.2 million
    — Dept. of Homeland Security – Customs and Border Protection, $254.2 million

Appendix C

Biographical Information: Committee on U.S. Army Corps of Engineers Science, Engineering, and Planning

David A. Dzombak (NAE), *Chair*, is the Walter J. Blenko, Sr. university professor of environmental engineering at Carnegie Mellon University, and faculty director of the Carnegie Mellon Steinbrenner Institute for Environmental Education and Research. He conducts research in water quality engineering and science, on topics pertaining to environmental restoration and the water-energy nexus. Dr. Dzombak is a member of the National Academy of Engineering, a registered Professional Engineer in Pennsylvania, a Diplomate of the American Academy of Environmental Engineers, and a Fellow of the American Society of Civil Engineers. He served as the chairman of the NRC *Committee on the Mississippi River and the Clean Water Act*. Dr. Dzombak holds a B.A. degree in mathematics from Saint Vincent College, B.S. and M.S. degrees in civil engineering from Carnegie Mellon University, and a Ph.D. degree in civil engineering from the Massachusetts Institute of Technology.

Patrick A. Atkins is with Atkins 360, LLC, a consulting firm in Pittsburgh that specializes in teaching and consulting on energy, lifecycle, and sustainability issues. Until his retirement in April 2007, Dr. Atkins was Director of Energy Innovation at Alcoa, Inc., responsible for implementing solutions for waste heat recovery in refining, smelting, and casting, assessment of alternate energy sources (e.g., renewable) and their applicability across Alcoa worldwide. Dr. Atkins is a member of the American Society of Civil Engineers, the National Society of Professional Engineers, and the Engineering Society of Western Pennsylvania. He is a past member of the Science Advisory Board for the Strategic Environmental Research and Development Program, and an Operating Advisor at Pegasus Capital Advisors, LLC, a New York-based Private Equity firm. Dr. Atkins is a registered professional engineer in Pennsylvania and Texas. Dr. Atkins received a B.S. degree in civil engineering from the University of Kentucky and his M.S. and Ph.D. degrees in environmental engineering from Stanford University.

Gregory B. Baecher (NAE) is the G.L. Martin Institute Professor of Engineering at the University of Maryland in College Park. Dr. Baecher's principal area of work addresses the reliability of civil infrastructure and project risk management, especially in geotechnical and water resources engineering. From 1998-2005 he served as a member of the NRC Water Science and Technology Board. Dr. Baecher is a member of the National Academy of Engineering and has served on several NRC committees. He chaired the NRC *Committee on Risk-Based Analyses for Flood Damage Reduction* and the *Panel on (Corps of Engineers) Methods and Techniques of Project Analysis*. He holds a B.S. degree in civil engineering from the University of California, Berkeley, and Sc.M. and Ph.D. degrees from the Massachusetts Institute of Technology.

Linda K. Blum is a research associate professor in the Department of Environmental Sciences at the University of Virginia. Her current research projects include study of how living organisms modify the geomorphology of salt marshes in response to external drivers such as sea level, precipitation, tides, and/or anthropogenic nitrogen loading; mechanisms controlling bacterial community abundance, productivity, and structure in tidal marsh creeks and soils; and rhizosphere effects on organic matter decay in anaerobic sediments. Dr. Blum served as chair of the NRC *Panel to Review the Critical Ecosystem Initiative* and was a member of the NRC *Committee on Restoration of the Greater Everglades Ecosystem*, the *Committee on Independent Scientific Review of Everglades Restoration Progress*, and the *Committee on Challenges and Opportunities in Earth Surface Processes.* She earned her B.S. and M.S. degrees in forestry from Michigan Technological University and her Ph.D. degree in soil science and microbial ecology from Cornell University.

Robert A. Dalrymple (NAE) is the Willard and Lillian Hackerman Professor of Civil Engineering at Johns Hopkins University in Baltimore, Maryland. His major research interests and projects are in the areas of coastal engineering, wave mechanics, fluid mechanics, littoral processes, and tidal inlets. His current interests are water wave modeling, tsunamis and their impacts on shorelines, and the interaction of water waves with the sea bed, specifically mud bottoms. He chaired the NRC *Committee on the Review of the Louisiana Coastal Protection and Restoration (LACPR) Program*, and currently is chairing the NRC *Committee on Sea Level Rise in California, Oregon, and Washington*. Dr. Dalrymple received his A.B. degree in engineering sciences from Dartmouth University, his M.S. degree in ocean engineering from the University of Hawaii, and his Ph.D. degree in civil and coastal engineering from the University of Florida.

Misganaw Demissie is director of the Illinois State Water Survey at the University of Illinois. His research at the Water Survey has focused on watershed science with emphasis on erosion and sedimentation and watershed hydrology. He has published more than one hundred journal articles, reports, and conference proceedings. Dr. Demissie is recipient of several awards including The Frank Bellrose Illinois River Conservation Award from the Nature Conservancy for outstanding service and contribution towards the restoration of the Illinois River. Dr. Demissie is a registered Professional Engineer in Illinois. He is a Fellow of the American Society of Civil Engineers, a Diplomate of the American Academy of Water Resources Engineers, and a member of the International Water Resources Association and the International Association of Hydrological Sciences. Dr. Demissie received his B.S. degree in civil engineering from the University of Iowa, and his M.S. and Ph.D. degrees in civil engineering from the University of Illinois.

Terrance (Terry) Fulp is the Deputy Regional Director for the U.S. Bureau of Reclamation's Lower Colorado River region headquartered in Boulder City, Nevada. Dr. Fulp is involved in numerous Colorado River issues, working with federal and state agencies and other stakeholder groups on system operations decisions. Prior to his appointment as Deputy Regional Director, he served as Area Manager of the Boulder Canyon Operations Office, where he managed a basin-wide effort to develop additional operational guidelines for Lake Powell and Lake Mead to minimize the effects of long-term drought. He was the principal investigator for the Department of the Interior's Watershed and River Systems Management Program. The program developed decision support tools for watershed management and resulted in development of RiverWareTM, a river operations modeling framework now is used by several water management agencies—including Reclamation and the Corps of Engineers. Dr. Fulp received his B.S. degree in earth sciences from the University of Tulsa, his M.S. degree in geophysics from Stanford University, and his Ph.D. in Mathematical and Computer Sciences from the Colorado School of Mines.

Larry Larson is the Executive Director of the Association of State Floodplain Managers (ASFPM), headquartered in Madison, Wisconsin. A founding member of ASFPM in the 1970s, Larson oversees the Association's activities and communications and coordinates national flood and water resources policy development and advancement with state, local, and federal agencies; the Administration and Congress; and other policy groups and organizations. He also spent 30 years with the Wisconsin Department of Natural Resources managing flood loss reduction, dam safety, and wetlands programs, and 5 years with the California Department of Water Resources on design and construction of large dams, aqueducts, and water projects. An expert in developing the nation's policy on wise and sustainable use of floodplains, Larry has authored numerous position papers and articles, provides expert testimony to Congress, and frequently speaks to policy makers and flood hazard managers nationally and abroad. He is a Certified Floodplain Manager and a registered professional engineer in California and Wisconsin. Larry holds a B.S. degree in civil engineering from the University of Wisconsin.

Diane M. McKnight is a professor in the Department of Civil, Environmental, and Architectural Engineering, and a fellow of the Institute of Arctic and Alpine Research, at the University of Colorado. Prior to her current post she was a research scientist with the U.S. Geological Survey, Water Resources Division. Her areas of research are biogeochemical processes, aquatic ecology, and reactive solute transport in streams and lakes in the Rocky Mountains and in polar desert areas of Antarctica. She has published numerous journal articles and book chapters, and edited several books. In 1995, along with three other limnologists, she co-authored *The Freshwater Imperative: A Research Agenda*. Dr. McKnight is past president of the American Society of Limnology and Oceanography and of the biogeosciences section of the American Geophysical Union, and was the first editor of the *Journal of Geophysical Research-Biogeosciences*. Since serving on the NRC *Committee on Climate Change and Water Resources Management* in 1990-1992, she has served on several other NRC committees, as well as the Polar Research Board and the Water Science and Technology Board. Dr. McKnight received her B.S. degree in mechanical engineering, M.S. degree in civil engineering, and Ph.D. degree in environmental engineering from the Massachusetts Institute of Technology.

J. Walter Milon is the Chair and Provost's Distinguished Research Professor in the Department of Economics, College of Business Administration, at the University of Central Florida. His major research interests are water resource economics, ecosystem valuation, and environmental policy. In addition to his academic research and publications, Dr. Milon has conducted research and consulting for a number of federal agencies including the Environmental Protection Agency, the National Marine Fisheries Service, and the National Oceanic and Atmospheric Administration. Dr. Milon received his B.S. degree in finance, and his M.S. and Ph.D. degrees in economics, from Florida State University.

A. Dan Tarlock is a professor at the Chicago-Kent College of Law, where he teaches courses in land use, property, energy and natural resource law, environmental policy, and international environmental law. He is an internationally recognized expert in environmental law and the law of land and water use. He has published a treatise, Law of Water Rights and Resources, and is a co-author of four casebooks. Professor Tarlock is a frequent consultant to local, state, federal, and international agencies, private groups, and law firms, and is an elected member of the American Law Institute. From 1989 to 1992, he served as chairman of the NRC *Committee on Western Water Management*. In 1996-1997 he was the principal report writer for the Western Water Policy Review Advisory Committee. He is a member of the California bar. Professor Tarlock is currently one of three United States special legal advisers to the NAFTA Commission on Environmental Cooperation. He also is a National Associate of The National Academies. Professor Tarlock received his B.A. and J.D. degrees from Stanford University.

Peter R. Wilcock is a professor of geography and environmental engineering at the Whiting School of Engineering at Johns Hopkins University in Baltimore, Maryland. His areas of research focus on river sedimentation processes and their role in stream restoration and river management. His research includes both laboratory and field experiments in sediment transport, open-channel flow, fluvial and hillslope geomorphology. Dr. Wilcock served as chair of the NRC *Panel on River Basin Systems and Coastal Planning* and was a member of the NRC *Committee on Grand Canyon Monitoring and Research*. He received his B.S. degree in physical geography from the University of Illinois, his M.S. degree in geomorphology from McGill University, and his Sc.D. degree from the Massachusetts Institute of Technology.